How To Start a Tree Nursery

I0446270

Step-By-Step Comprehensive Blueprint on How to Start a Tree Nursery Business and Thrive

Introduction

- ❖ *Are you passionate about trees and the environment?*

- ❖ *Do you dream of turning your love for nature into a thriving business?*

- ❖ *Have you ever wondered what it takes to start and run a successful tree nursery?*

If you've answered YES to any of the questions above, this book has the answers you seek!

This book will introduce you to the fascinating world of tree nurseries and provide valuable insights, practical advice, and a step-by-step success blueprint.

Whether you're a seasoned entrepreneur or a nature enthusiast looking for a new venture, this comprehensive guide will equip you with the knowledge and tools you need to thrive in the tree nursery business.

In it, you'll discover things like:

- The basics of the tree nursery business

- How to select the right location for your tree nursery

- Where to source quality seeds and seedlings

- How to take care of your seedlings and nutrition

- Different tree species and market demand

- Different marketing strategies for tree nurseries

- How to create ideal growing conditions

- And so much more!

After reading this book, you will have the knowledge, skills, and confidence to establish and grow a successful tree nursery.

Whether you dream of a small, intimate nursery or aspire to become a major supplier, "How to Start a tree nursery" will be your trusted guide as you embark on this rewarding journey.

So, are you ready to dive into the world of tree nurseries?

Let's begin our exploration!

PS: I'd like your feedback. If you are happy with this book, please leave a review on Amazon.

Please leave a review for this book on Amazon by visiting the page below:

https://amzn.to/2VMR5qr

Table of Contents

Chapter 6: Pricing Your Plants for Success

Conclusion

Chapter 1: Starting a Tree Nursery - Cultivating Your Green Business Dream

Statistics reveal a soaring demand for trees.

Projections show that the global tree nursery industry will continue growing, with an annual growth rate of 8%. This remarkable growth comes from the increasing need for green spaces, landscaping projects, and ecological restoration initiatives.

This means the opportunity for breaking into the industry is still very ripe, even for beginners.

This chapter will introduce you to the exciting world of starting a tree nursery business. We will discuss critical processes, considerations, and crucial factors in turning your green thumb interest into a thriving business.

Whether you are a seasoned horticulturist or a nature enthusiast with limited expertise, this chapter will equip you with the knowledge and direction to help you confidently and successfully build your tree nursery.

Without further ado, let's jump straight into it.

Step 1: Finding the Perfect Land for Your Garden

The first thing you have to do is find an appropriate piece of land for your nursery.

Before evaluating any factors, you should consider the land size and availability you are about to find out. If you don't own the land, you can buy or explore alternative options such as leasing, collaboration, or utilizing community garden spaces.

Calculate the land size you'll need to accommodate all aspects of your nursery operations, including growing areas for your trees, infrastructure such as propagation areas and potting zones, storage facilities, and administrative offices.

Also, account for the layout and design of your nursery, and ensure there is enough space for efficient workflow and movement of equipment and materials.

Assess whether you have suitable land in your desired location to accommodate your nursery operations. Therefore, begin evaluating the availability of land in the location you have in mind. Determine if there are suitable and accessible land parcels that meet your specific needs. Other factors you should consider that we'll cover are land ownership, zoning regulations, and any potential restrictions that may affect your nursery operations.

After finding land that fits your desired size, the next step is to evaluate whether it fits the needs of a tree garden. This means you must ensure the space has suitable soil quality, drainage, sunlight exposure, and access to water sources.

The right location provides favorable conditions for optimal growth and development of your seedlings.

How to Choose the Right Location for Your Tree Nursery

Here are some factors to consider when selecting the ideal site:

1: Climate Considerations

Are the climatic conditions in your area going to be favorable?

The climate plays a pivotal role in the health and growth of your nursery trees. Understanding your local climate patterns and their effects on tree species is paramount. Consider factors such as temperature ranges, frost dates, rainfall patterns, and prevailing winds.

For example, the US has a range of climate zones, from cold temperate zones in the north to subtropical and arid zones in the south. Research the specific climate zone of your chosen location and understand the average temperature ranges, frost dates, and growing seasons. Doing this will help you choose the right tree for your tree nursery. Choose tree species well-suited to your climate zone. For instance, native or adapted species, i.e., oak and pine trees, tend to thrive better and require less maintenance.

Also, be aware of microclimates within your nursery location. Certain areas of your nursery might be warmer or cooler than the surrounding environment due to various factors, like shading, wind patterns, and topography. Plan your layout accordingly to optimize plant growth.

- Understand your nursery's sunlight patterns. By strategically planting tree species according to their sunlight needs, you will have healthy, attractive stock that appeals to customers and thrives in various environments.

- Choose a location with ample sunlight for a successful tree nursery. Arrange taller plants to provide shade for those that prefer less direct sunlight. Group sun-loving plants in areas that receive ample sunlight, while shade-tolerant ones can find their homes under tree canopies or structures.

- Factor in prevailing winds to shield delicate plants from potential damage. Position windbreaks like fences or larger plants on the windward side to create a microclimate that reduces stress on your green residents.

Invest in protective measures if your nursery is in an area prone to extreme weather events such as frost, heavy snowfall, hurricanes, or drought. This could include greenhouses, windbreaks, frost cloth, or irrigation systems to regulate moisture during dry spells.

Nursery owner's tips:

- ➤ Regularly monitor weather patterns and conditions in your nursery. Doing this will allow you to adjust irrigation schedules, implement protective measures, and make informed decisions about planting and care.

- ➤ Seek advice from local agricultural extension offices, horticultural experts, and other tree nurseries in your area. These experts can provide valuable insights into your region's climate challenges and solutions.

> ➤ Familiarize yourself with the USDA Plant Hardiness Zone Map, which divides the US into zones based on average annual minimum temperatures. This map helps you choose tree species suited to your specific zone and minimize the risk of frost damage.

Closely linked to climate is access to water.

2: Access to Water

Is there enough water to cater to your tree's demand?

Your nursery trees need sufficient water to stay hydrated and healthy. You ought to look for a location that provides access to a reliable water source, such as wells, ponds, rivers, or the local municipal water supply. Having a dependable water supply will help you meet the watering requirements of your trees, especially during dry periods.

Don't stop there; also:

- **Consider water rights:** If you plan to source water from wells, rivers, or other natural sources, please consider the availability of water rights. Ensure you have the necessary permits and legal permissions to use water from these sources. Familiarize yourself with local regulations and consult relevant authorities to ensure compliance.

- **Water quality:** Along with availability, water quality is a critical aspect to consider. Assess the quality of the water source in terms of its purity and suitability for irrigation purposes. If the water contains high salinity levels, contaminants, or other impurities, it may negatively affect your trees' health and growth. Water testing can provide valuable insights into its suitability for your nursery needs.

- **Installation of irrigation systems**: Evaluate the feasibility of installing irrigation systems to distribute water throughout your nursery effectively. Depending on your nursery's size and layout, consider options like drip irrigation, sprinkler systems, or other efficient irrigation methods that can help ensure your trees receive the necessary water in a controlled and efficient manner.

3: Accessibility and Logistics

On accessibility and logistics, you'll want to assess how easy it is to reach your chosen location and consider its proximity to transportation routes.

How so?

The location you choose for your tree nursery should be easy to access for transportation purposes. Consider whether delivery trucks can easily reach your nursery without facing obstacles or difficult terrain. You want to ensure a smooth flow of incoming and outgoing shipments, making it convenient for suppliers and customers.

Another consideration is whether the location offers enough space and flexibility to accommodate your growth plans. Aim to have room for increased production capacity, adding new facilities, or diversifying your tree varieties as your business expands.

Select a location that allows for future growth and increased production capacity. Assess the zoning regulations and land-use restrictions to ensure your nursery can grow and evolve within the established guidelines.

Also, consider the availability of additional land for expanding your nursery operations or introducing new tree varieties.

Assess the zoning regulations and land-use restrictions to ensure your nursery can grow and evolve within the established guidelines.

- **Local support and infrastructure:** When considering local support and infrastructure, assess the availability of nearby services that can contribute to the success of your tree nursery. Look for areas with nearby suppliers of nursery materials, such as seeds, fertilizers, potting soil, and other essential supplies. Local suppliers can provide convenience and reduce transportation costs, ensuring a steady and efficient supply of necessary materials for your nursery operations.

- In addition, consider the availability of skilled labor in the vicinity of your chosen location. Skilled workers experienced in horticulture, tree propagation, and nursery management can contribute to the smooth functioning of your nursery.

- Look for areas offering a pool of experienced workers or potential training opportunities to develop the necessary skills. Finally, look for locations near research institutions or agricultural extension services. These institutions can provide valuable guidance, support, and access to research findings relevant to tree nursery practices. They may offer educational programs, workshops, or consultations that can help enhance your nursery operations and keep you updated on the latest industry trends.

4: Soil quality

Is the soil suitable for growth?

Soil quality is a foundational factor when considering the location of your tree nursery. It plays a crucial role in the health and growth of your plants, and choosing the right soil can significantly impact the success of your nursery business.

Here's why soil quality matters and how it influences your nursery's location decision:

- **Plant health and growth:** The soil type in a specific location affects its drainage, water-holding capacity, and nutrient content. Different plant species have varying soil preferences. For instance, trees that prefer well-draining soil may struggle in areas with poor drainage, leading to root rot and stunted growth. On the other hand, trees that thrive in moist conditions may struggle in sandy or rocky soils. By understanding the soil types in your potential locations, you can ensure your chosen site is compatible with the needs of your tree species.

- **Nutrient availability:** Soil serves as a reservoir for essential nutrients that plants need for healthy growth. Nutrient-rich soil fosters strong root development, lush foliage, and optimal overall plant health. Before selecting a location, assess the soil's nutrient content and ensure it aligns with the requirements of the trees you intend to grow. Soil testing can help determine whether the soil needs any amendments to provide essential nutrients.

- **Water management:** Soil type significantly influences water absorption and retention. Soils with good water-holding capacity can support moisture-loving trees without the risk of overwatering while well-draining soils prevent waterlogged conditions that can lead to root diseases. Understanding the drainage characteristics of the soil in potential locations will help you manage irrigation and preventing water-related issues.

- **Soil preparation and amendments:** Different soil types may require various amendments to create optimal growing conditions. For instance, adding organic matter to a sandy soil can improve water retention, while adding drainage materials to heavy clay soil can enhance drainage. Evaluating the soil quality in potential locations allows you to plan for necessary soil amendments and preparation techniques.

- **Environmental sustainability:** Choosing a location with soil that matches the preferences of your tree species can reduce the need for excessive irrigation, fertilization, and pest control measures. This can contribute to a more environmentally sustainable nursery operation by minimizing resource use and reducing the potential for soil and water contamination.

In conclusion, soil quality is a fundamental consideration when choosing a suitable location for your tree nursery. It directly impacts plant health, growth, and overall success. By selecting a location with soil that aligns with the needs of your tree species and by implementing appropriate soil management practices, you can create a thriving and sustainable nursery that produces healthy, vibrant trees.

Now that you've got the perfect spot, you must also ensure your nursery meets the area's local regulations. As a result, be sure to obtain the required licenses.

What are some of these licenses and regulations? Let's find out!

Chapter 2: Establishing Your Tree Nursery Business

Starting a tree nursery is not just about cultivating beautiful and healthy trees. It also involves navigating various regulations and permits to establish a legitimate and thriving enterprise.

NOTE: Although different laws apply in different countries, this book will use the US regulatory guidelines. However, remember that the advice in this book does not constitute legal advice.

Therefore, please consider working closely with a qualified attorney or other qualified professional to navigate this process to ensure you stay within the legal boundaries. Complying with federal, state, and local regulations ensures your tree nursery operates legally and without disruptions.

This chapter will explore the key regulatory issues you must address before starting your tree nursery in the US. By the end of this chapter, you will have a comprehensive understanding of the legal requirements and considerations involved in establishing your tree nursery business.

With that mentioned, let's sail and explore these legal requirements and considerations you'll need to establish your tree nursery business.

1: Business Registration and Licensing

The first step in establishing your tree nursery is registering your business with the appropriate authorities. Decide on a suitable business structure, such as sole proprietorship, partnership, limited liability company (LLC), or corporation, and register it with the Secretary of State's office.

Choosing the appropriate business structure is not something you can overlook because it affects your liability, taxes, and how you manage your business.

How so?

Below are some details about each business structure:

- **Sole Proprietorship:** This is the most basic type of business ownership where you run the nursery as an individual. However, remember that in a sole proprietorship, your assets are not separate from the business, which means you have unlimited personal liability for the nursery's debts and obligations. In other words, as the sole proprietor of the nursery, you are personally accountable for everything from debts, obligations, and legal liabilities incurred by the business.

- **Partnership:** In a general partnership, partners share profits and liabilities equally. A limited partnership has general participants who have personal liability and limited partners who have limited liability. If your intention is to start the nursery with someone else, you may opt for a partnership.

- **Limited Liability Company (LLC):** This business structure combines the limited liability of a corporation with the flexibility and tax advantages of partnerships. An LLC generally protects your assets from business debts and lawsuits.

- **Corporation:** A company is a distinct legal entity from the stockholders who are its owners. It protects its stockholders from limited liability to its shareholders but involves more complex legal and tax requirements.

For a beginner starting a tree nursery, the Limited Liability Company (LLC) structure is often the best option due to its combination of limited liability protection and flexibility. However, it's important to note that the best business structure for a tree nursery may vary depending on individual circumstances, future growth plans, and the specific goals of the owner(s).

NOTE: Please consult a business attorney or a qualified financial advisor who can help you fully understand the legal and tax implications. These professionals can provide personalized advice based on your unique situation and help you make an informed decision that aligns with your business objectives and risk tolerance.

2: State Registration

After choosing a business structure, the next step is to register your tree nursery with the appropriate state authorities. This step involves filing the necessary documents with the Secretary of State's office. The registration process and requirements vary from state to state, so research your state's specific procedures and fees.

3: Employer Identification Number (EIN)

You must also obtain an EIN from the Internal Revenue Service (IRS). This is crucial for tax purposes, especially if you plan to hire employees. An EIN is also required if you operate as an LLC or corporation. It is essentially a unique identifier for your business, similar to a Social Security Number for individuals.

4: Business Licenses and Permits

Check with your local municipality or county to determine if you need additional business licenses or permits to operate a tree nursery. The requirements may vary based on your location and the specific activities of your nursery.

Common licenses you might need include a nursery license, general business license, environmental permit, zoning permit, Water Use Permit, Sales Tax Permit, Transportation Permit, Occupational Safety and Health Administration (OSHA) Compliance, Signage Permit, or agricultural permit.

Let's explain these:

- **Nursery License:** This specialized license is specific to businesses that propagate, grow, and sell plants.

- **General Business License:** This is a basic license required to operate any business in your area.

- **Agricultural Permit:** You may need this permit if your nursery will engage in large-scale agricultural activities, such as growing crops.

- **Zoning Permit:** This permit ensures your nursery operation complies with local zoning regulations.

- **Water Use Permit:** Required if your nursery uses significant water quantities for irrigation or other purposes.

- **Sales Tax Permit:** If you plan to sell plants directly to customers, you'll likely need this permit for collecting and remitting sales tax.

- **Transportation Permits:** You might need transportation permits if you intend to transport plants or products across state lines.

- **Occupational Safety and Health Administration (OSHA) Compliance:** Ensures workplace safety standards for your employees.

- **Signage Permits:** Required if you plan to use sign advertising.

5: Compliance with Local Regulations

Beyond business registration and licensing, be aware of local regulations or zoning ordinances that could affect your tree nursery. Some areas may have restrictions on the size of the nursery, the location of structures, or the use of certain chemicals. Ensure your nursery follows all local regulations and restrictions.

6: Zoning and Land Use Regulations

Before selecting a site for your tree nursery, review local zoning regulations to ensure local authorities have zoned the area for agricultural or nursery purposes. Different regions have specific land-use restrictions; check these to confirm that your nursery business complies with all zoning requirements.

Common zoning classifications suitable for a tree nursery include agricultural, rural residential, or commercial zoning. Sometimes, you may need to apply for a conditional use permit or seek a variance if local authorities have not directly zoned your intended location for nursery operations.

Note that zoning regulations can vary significantly between jurisdictions, so research and understand the specific regulations that apply to your intended location.

7: Environmental Permits and Conservation Compliance

Tree nurseries often deal with environmental considerations, especially concerning water usage and potential impacts on natural habitats. Depending on the scale of your nursery and its proximity to wetlands, water bodies, or protected areas, you may need to obtain permits related to water usage, wetland impacts, or environmental conservation compliance. Consult local environmental agencies and the U.S. Army Corps of Engineers to ensure you meet all the requirements.

Also, some zoning regulations may address environmental concerns, such as wetlands, floodplains, or protected habitats. Verify whether your intended location has any ecological restrictions that could affect your nursery's operations. You may need to obtain additional permits or cater for environmental impact assessments if your nursery is near sensitive areas.

8: Plant Health Regulations

The US Department of Agriculture Animal and Plant Health Inspection Service (APHIS) enforces regulations to prevent the spread of pests and diseases through nursery stock.

Depending on the type of tree species you plan to grow and sell, you may need to comply with specific phytosanitary regulations and obtain the necessary plant health certificates. This ensures your nursery stock meets the required standards and is no threat to the local environment or agriculture.

9: Sales Tax and Nursery Certificates

Determine the sales tax requirements in your state for nursery products. In some states, sales tax may not apply to certain agricultural products, including nursery stock. Obtain a nursery certificate or tax exemption, where applicable, which allows you to purchase supplies and materials without paying sales tax.

10: Intellectual Property and Trademarks

If you have developed or acquired unique tree cultivars or varieties, consider protecting your intellectual property through patents or trademarks. This can prevent other nursery owners from reproducing or selling your proprietary plant material without your permission.

11: Labeling and Marketing Regulations

Labeling and marketing regulations are a key part of the tree nursery business because they ensure transparency and consumer protection. When selling trees, it's essential to provide accurate and clear labeling on each plant to inform buyers about the species, characteristics, and potential risks associated with the tree.

Accurate labeling helps customers make informed decisions, especially when choosing the right trees for their needs and locations. Key information to include on the labels may include:

- **Species Information:** Clearly state the scientific and common name of the tree species. Doing this helps customers identify the exact type of tree they are purchasing and ensures they receive the desired characteristics and benefits.

- **Growth and Care Instructions:** Provide essential information about the tree's growth requirements, including sunlight exposure, water needs, soil preferences, and maintenance tips. Customers will appreciate guidance on nurturing their new trees to ensure successful growth.

- **Size and Age:** Indicate tree's size and age to ensure customers have a clear understanding of its current maturity and potential future growth.

- **Potential Risks or Warnings:** If certain tree species have specific hazards, such as thorns, allergens, or toxic properties, include appropriate warnings to help customers take necessary precautions and make informed choices.

- **Origin and Sourcing:** If your tree nursery promotes locally sourced or eco-friendly practices, consider highlighting this information on the labels. Consumers often appreciate products with transparent and sustainable sourcing.

- **Certifications or Ratings:** If your tree nursery has received certifications for organic practices, sustainable sourcing, or industry quality standards, display these achievements on the labels to build customer trust.

- **Contact Information:** Include the nursery's contact details on the label, such as a phone number or website, for customers to reach out for further inquiries or support.

Adhering to labeling and marketing regulations can help you demonstrate commitment to consumer protection, build trust with your customers, and showcase your dedication to providing quality trees. Compliance with these regulations fosters customer satisfaction and ensures your tree nursery operates ethically and responsibly within legal bounds.

Always stay updated on relevant local and national regulations to maintain accurate and informative labeling practices throughout your tree nursery business.

Tree nursery owner's tips:

- ❖ Establishing a tree nursery in the United States involves navigating through various regulatory requirements. Therefore, it's important to remember that many of these steps are essential to ensuring your business operates legally and sucessfully. While these regulations may initially seem overwhelming, taking the time to address these permits and regulations will ultimately benefit your nursery in the long run.

- ❖ While there are expenses associated with regulatory compliance, it's important to view them as necessary investments in the growth and sustainability of your tree nursery business. Budgeting for these costs and planning accordingly will help you manage expenses effectively. Additionally, consider seeking guidance from business advisors or experts who can help you navigate the process efficiently and potentially find cost-saving strategies. Some local or regional agencies may offer support or incentives for certain agricultural businesses, including nurseries.

- ❖ Do not forget that investing in regulatory compliance and proper business setup is an important step toward building a successful and reputable tree nursery. The benefits of operating legally, maintaining a positive reputation, and adhering to industry standards will ultimately contribute to your overal profitability and business growth.

After thoroughly considering the factors involved in choosing the right location for your tree nursery and the licenses required, the next step is to focus on building your stock and preparing for nursery operations.

Chapter 3: Build Your Stock and Prepare for Tree Nursery Operations

After settling for a location, the next thing you need to know about is how to build your stock.

Building your stock and preparing for nursery operations involves acquiring the necessary tree seedlings, seeds, or young trees and getting your nursery ready for planting and growing activities. This phase includes several steps to ensure you obtain a healthy and diverse stock of trees and your nursery has everything it needs to support its growth.

Below are some key components you need to consider:

Choosing the Right Trees

When it comes to tree selection and procurement for your tree nursery, everything comes down to taking your time to research and determine the tree species or varieties you want to cultivate in your tree nursery.

Consider both native and non-native species. Native trees offer benefits like adaptability to local conditions and support for local wildlife. Non-native species can provide diversity and unique aesthetics. Specialty trees, such as dwarf varieties, ornamentals, and bonsai, cater to niche markets and passionate collectors.

As someone starting a tree nursery business, it is wise to take a balanced approach that includes a mix of native and non-native species and specialty trees.

How so?

Let's say you decide to start a tree nursery business in the Midwest region of the United States, which is known for its varying climate and rich biodiversity. Here's how you can strategically select tree varieties to cater to different customer preferences:

Let's explore each category and why they could be beneficial:

Native Trees for Sustainability:

The importance of native trees is ecological balance and local wildlife support. Trees species such as Red Maple (Acer rubrum), White Oak (Quercus alba), and Eastern Redbud (Cercis canadensis) are great examples of native tree species you can find. These trees are well-adapted to the Midwest's climate and offer environmental benefits. These trees would support local ecology and appeal to environmentally conscious customers.

For ornamental Non-Native Trees:

You can attract homeowners by introducing non-native ornamental trees like Japanese Maple (Acer palmatum) and Flowering Dogwood (Cornus florida). These trees offer unique foliage colors and attractive flowers, making them popular among aesthetic-focused customers.

Drought-Tolerant Varieties:

Given the region's (Midwest region) occasional dry spells, you can also have drought-tolerant species such as Prairie Crabapple (Malus ioensis) and Colorado Blue Spruce (Picea pungens).

Once established, these trees require less watering and resonate with environmentally conscious customers seeking low-maintenance options.

Specialty Trees for Enthusiasts:

You can cater to gardening enthusiasts by offering specialty trees like Dwarf Hinoki Cypress (Chamaecyparis obtusa), Japanese Black Pine (Pinus thunbergii), and Trident Maple Bonsai (Acer Buergerianum). These trees appeal to collectors, hobbyists, and those who appreciate the art of bonsai cultivation.

By offering a balanced selection of native, non-native, and specialty trees, your nursery could cater to a wide customer base while capitalizing on niche markets. This approach would create a diverse and appealing inventory and establish the nursery's reputation as a comprehensive and innovative source that suits various landscaping needs and preferences.

But even so, please also consider factors like market demand, local preferences, and the suitability of the chosen species to your specific climate and region. Why is that the case? This is because you want to select tree species that have a high demand in your target market and are well-adapted to thrive in the local environmental conditions.

Research Market Demand

Please also explore the market demand for the tree species you want to grow. Where do you start?

Start by exploring online resources like industry reports or studies related to tree nurseries. Look for information about popular tree species, emerging trends, and the needs of potential buyers. These reports often contain valuable data on market trends, demand for specific tree species, and predictions for future growth.

You can also browse online marketplaces that sell plants and trees, like Etsy, Amazon, and plantz.com[1], which offer a wide variety of plants and trees for sale. Observe which tree species are frequently searched for or have positive reviews because this indicates their popularity among customers.

Another way to do it is to read articles and blogs from reputable online gardening magazines and blogs. These sources often feature information on trending tree species and gardening practices. Alternatively, you can watch gardening-related videos on platforms like YouTube. Many gardening enthusiasts and experts share their experiences and recommendations, which can help you understand customer preferences.

Networking is another way to do it by linking up with other tree nursery owners, landscapers, or gardening enthusiasts in your area.

Finally, check local government departments or environmental agencies about tree planting initiatives or reforestation projects. These efforts indicate potential demand for specific tree species. Remember to take notes and organize the information you gather. Pay attention to recurring themes and preferences mentioned by different

---

[1] https://www.plantz.com/

sources. This comprehensive online research will serve as the foundation for your tree nursery's stock selection and ensure you grow and offer tree species that are in demand and align with the interests of potential buyers.

Explore Reputable Nurseries, Suppliers, or Seed Banks

After settling on tree species, it's time to explore reputable sources for procuring high-quality seeds, seedlings, or young trees. Look for reputable nurseries, suppliers, or seed banks specializing in the tree species you are interested in. Check their track record, customer reviews, and certifications to ensure they provide reliable and top-notch stock.

Certifications such as ISO, organic certifications, or certifications from horticultural organizations indicate a commitment to quality and adherence to industry standards. These certifications can give you assurance about the reliability and integrity of their tree stock.

After narrowing your list, consider directly visiting or contacting potential sources. This lets you get more detailed information about their products, services, and growing practices. Ask questions about their propagation methods, seed collection processes, and any guarantees or warranties they offer for their tree stock.

TIP: If possible, request samples or place small orders to assess the quality of the tree stock firsthand. This can help you evaluate the health, vigor, and adaptability of their seedlings or young trees. Pay attention to factors such as root development, leaf health, and overall appearance to ensure you are getting high-quality stock.

The next step is designing propagating areas; you'll need these important areas within your nursery for propagation activities such as budding/grafting or sowing. But before that, you need to craft the optimal garden layout for your tree nursery. This will help you design your garden for optimal performance.

How can you achieve this?

Creating the Optimal Tree Nursery Layout

Aim to understand the water and light requirements of the different tree species you intend to grow in your nursery. Categorize them based on how much water and sunlight they need to thrive. Organize trees based on their water and light preferences. This strategic approach optimizes your tree nursery for better tree health, growth, and maintenance.

For instance, you can group drought-tolerant tree species and create zones for shade-loving trees. This will allow you to create micro-environments that mirror the natural conditions these trees thrive in. With those grouped trees in mind, you can then strategically position irrigation systems, shading structures, and windbreaks to cater to each group's unique requirements. This will save water and energy and reduce the risk of overexposing certain trees to harsh elements.

With a grouped layout, caring for the trees will become more efficient. You or your staff will manage to water, fertilize, and systematically monitor trees, resulting in healthier, happier trees and a reduced workload for you and your team.

- You can strategically arrange trees with taller species toward the back and shorter ones in the front. This ensures each tree receives adequate sunlight.

- Install informative signage highlighting each species' unique features and care requirements. This adds educational value to the visitor experience.

- Focus on visual appeal by considering color contrasts, complementary foliage textures, and seasonal blooms.

To illustrate: Strategically introduce flowering trees that blossom during different seasons. This dynamic display creates an ever-changing canvas of colors.

Propagation Areas Establishment

Set aside specific areas within your nursery for propagation activities. These areas will be for creating new tree seedlings through different methods.

These areas include:

Seed Sowing Area

A seed sowing area is one dedicated to planting tree seeds in propagation trays or pots filled with suitable potting soil. Seed sowing is a common way to start many new trees from seeds at once. You can easily acquire seed sowing trays and pots from various local garden supply stores, online gardening retailers (like Amazon, eBay, and specialized gardening websites, like Burpee[2] and Gardener's Supply Company,[3] which are reliable gardening websites, or your local specialty nursery equipment suppliers.

Regarding potting soil for your tree nursery, choosing the right soil type and quality is the best way to ensure the healthy growth of your seedlings and young trees.

---

[2] https://www.burpee.com/
[3] https://www.gardeners.com/

Here's what you need to know about obtaining potting soil:

1. **Garden Supply Stores:** Local garden centers and nurseries typically stock a variety of potting soils suitable for different plant types. Visit these stores to explore options and seek recommendations from knowledgeable staff.

2. **Home Improvement Stores:** Many home improvement stores carry a selection of potting soils, especially during gardening seasons. Check the gardening or outdoor section of these stores for suitable options.

3. **Online Retailers:** Platforms like Amazon, gardening websites, and specialty retailers offer various potting soils. Read product descriptions and customer reviews to ensure you're selecting a high-quality and appropriate soil mix.

4. **Nursery Suppliers:** Some suppliers specialize in providing horticultural supplies to nurseries. They often offer bulk quantities of potting soil suitable for commercial use. Contact these suppliers for specific recommendations and orders.

5. **Local farms or composting facilities:** Some local farms or composting facilities produce organic potting soil blends. These can be environmentally friendly options that support sustainability. So, you can also explore this option as well.

When choosing potting soil, consider all factors, like the type of trees you're growing, drainage properties, nutrient content, and whether you require specific soil mixes (e.g., for acid-loving plants). Look for potting soils that provide good aeration, water retention, and appropriate pH levels for your tree species.

Additionally, consider using potting soil blends enriched with organic matter and nutrients to support healthy root development and robust plant growth.

Several available and reputable options cater to different tree species and growing conditions. Here are some popular potting soil brands and blends to consider:

- **Miracle-Gro Potting Mix:** This widely recognized brand offers various potting soil mixes for different plants, including trees. They provide options for indoor and outdoor use and nutrient-enriched blends.[4]

- **Fox Farm Ocean Forest Potting Soil:** This organic potting mix is known for its nutrient-rich composition. It is rich in earthworm castings, bat guano, and aged forest products to support healthy plant growth.[5]

- **Espoma Organic Potting Mix:** Espoma offers organic potting soil blends enriched with beneficial fungi blends that promote root development. They also have specialized blends for specific plant types.[6]

- **Black Gold All-Purpose Potting Soil:** This brand provides a versatile potting soil mix suitable for various plants, including trees. It offers good drainage and aeration properties.[7]

- **Proven Winners Premium All-Purpose Potting Soil:** Proven Winners is known for its high-quality

---

[4] https://tinyurl.com/3vutn9bv

[5] https://foxfarm.com/product/ocean-forest-potting-soil

[6] https://www.espoma.com/product/espoma-organic-potting-mix/

[7] https://blackgold.bz/products/potting-mixes/all-purpose/

plant products, and their potting soil is no exception. This mix uses a formulation that is excellent at water retention and aeration.[8]

- **Happy Frog Potting Soil:** As another offering from Fox Farm, this potting mix has a design that creates a favorable environment for root growth. It contains beneficial nutrients and soil microbes to enhance plant health.[9]

When choosing a potting soil option, consider the specific needs of the tree species you're cultivating and your local climate and environmental conditions.

It's also helpful to read reviews, consult gardening forums, and seek recommendations from fellow tree nursery enthusiasts to find the best fit for your nursery's requirements.

---

[8] https://tinyurl.com/2c3khm64

[9] https://foxfarm.com/product/happy-frog-potting-soil

Figure 1: Young plant in a black plastic seedling tray

This will allow you to start growing your trees from their very early stages and nurture them into healthy young plants.

Grafting or Budding Station

This station will help you perform grafting or budding to propagate specific tree varieties.

Grafting involves joining a scion (a young shoot or bud) from one tree onto the rootstock of another, while budding involves inserting a bud from one tree onto the bark of another.

Cutting Propagation Zone

In this zone, you take cuttings from healthy branches of parent trees and plant them in the potting medium. Cutting propagation is another common way to produce new tree seedlings with the same characteristics as the parent tree.

Figure 2: Young cuttings in a propagation zone

Growing Beds or Containers Creation

It is important to have a secure space for nurturing these tree seedlings. Growing beds and container creation are what you need to provide a space where the tree seedlings will be planted and nurtured.

A growing bed is a designated area within a nursery where you cultivate plants directly in the ground. It's a prepared plot of soil that provides a suitable environment for plant growth. Growing beds are ideal for larger plants or trees requiring more space and root development.

On the other hand, a container refers to any receptacle used for holding and growing plants. Containers come in various sizes and materials, such as plastic pots, clay pots, fabric bags, or even specialized tree seedling trays. Containers are especially useful for starting plants from seeds or cuttings and growing smaller plants or seedlings. They allow for better control over soil quality, drainage, and overall growing conditions.

Figure 3: Example of a container creation

Figure 4: A tree nursery bed

Since you want to start a tree nursery, you must acquire growing beds & containers for your nursery operations. This involves researching and selecting suitable containers and growing beds that align with the types of trees you intend to cultivate. These containers could include pots, trays, or other specialized containers explicitly designed for tree seedlings and young plants.

You can purchase these items from garden supply stores, online retailers, or specialized nursery equipment suppliers. This equipment will help you provide proper growing conditions, organize the nursery space, and ensure successful seedling propagation and growth.

Here's a simplified approach to determining which container to use for your tree nursery:

- **Research tree species**: Understand the specific tree species you want to cultivate. Research their typical root depth, growth rate, and space requirements. Reputable gardening books, online resources, and local experts can provide this information.

- **Match container size:** It's generally better to choose a container at least twice the diameter of the tree's root ball to allow for adequate root growth and prevent the tree from becoming root-bound too quickly.

- **Consider depth:** Tree roots need space to grow downwards. Select containers that provide enough depth for root development. Deep containers also help prevent waterlogging.

- **Drainage:** Ensure the chosen container has proper drainage holes to prevent water accumulation and root rot. Ensure they have drainage holes at the bottom to allow excess water to escape.

- **Material matters:** Different materials have varying insulation properties. Consider using light-colored containers to reduce heat absorption for hot climates, while dark containers might suit colder regions.

- **Portability:** If you plan to move the trees around, opt for containers with built-in wheels or handles for easier transportation.

- **Transplanting plans:** Consider how long you intend to keep the tree in the container. A smaller container will suffice if the goal is to use it as a temporary stage before transplantation. If the tree will stay longer, choose a larger one to accommodate growth.

- **Review guidelines:** Some tree species have specific container sizes and type recommendations. Check if the species you're cultivating has any published guidelines.

Here is a simplified approach to determining which bed to use for your tree nursery:

- **Identify tree characteristics:** Understand the specific needs of the tree species you plan to grow. Consider factors like root depth, growth rate, and space requirements.

- **Match needs to bed type:** Choose a growing bed type that aligns with the identified characteristics. For instance, trees with deep root systems might need deeper raised beds, while slower-growing species might thrive in standard ground beds.

- **Consider drainage:** Ensure the chosen growing bed provides proper drainage to prevent waterlogging, which can harm tree roots. Use well-draining materials for bed construction, such as gravel or raised platforms, which allow excess water to escape.

- **Assess space availability:** Consider the available space in your nursery and arrange growing beds to maximize efficient land use.

- **Adaptability:** Stay open to adjustments based on your nursery's unique microclimates, soil conditions, and localized factors.

By considering these steps and aligning the characteristics of your trees with the appropriate growing bed type, you'll create an environment that fosters healthy growth and successful tree cultivation.

Remember, while these guidelines offer a starting point, individual circumstances might require you to adjust accordingly. It's important to remain flexible and adaptable enough to modify things based on the specific tree species' needs and nursery site conditions. This approach ensures your trees receive the optimal conditions for growth and sets the nursery up for success. Regularly monitoring your trees' growth and health will help you gradually fine-tune your container choices.

In addition to preparing for the growth of your tree seedlings, you need to focus on creating the perfect growing beds or containers where your young plants will thrive. This stage helps establish a strong foundation for healthy and robust tree growth.

Let's dive into the essential steps and considerations for this process:

Preparing Clean and Well-Drained Beds

The first step is ensuring the growing beds are clean and free from debris or previous plant residues. You want a fresh start for your tree seedlings! Also, ensure proper bed drainage to prevent soggy conditions and root rot.

TREE NURSERY OWNER'S TIP: If your nursery area is prone to heavy rainfall or poor drainage, consider using raised beds. Raised beds provide better control over soil moisture levels and minimize the risk of waterlogging, especially during rainy seasons.

However, set up your tree nursery in a location with good natural drainage to avoid water accumulation. Avoid areas prone to flooding or stagnant water, as they can jeopardize the health of your young trees

Figure 5: A tree nursery with raised garden beds

If you would rather use containers:

- Before using new containers or reusing old ones, thoroughly clean them to remove any dirt, debris, or previous plant residues. Use a mixture of water and mild soap or diluted bleach solution to sanitize the containers. Rinse them well with clean water to ensure no residue remains.

- Check each container for adequate drainage holes at the bottom. Proper drainage helps prevent waterlogging, which can harm the seedlings' roots. If the containers do not have drainage holes, use a drill or sharp tool to create small holes at the bottom.

- Place containers on raised platforms or pot feet to let excess water flow freely through the drainage holes. Elevating the containers ensures water does not accumulate beneath them, promoting better drainage. Use a mixture of water and mild soap or a dilute bleach solution to sanitize the containers. Rinse them well with clean water to ensure no residue remains.

- Regularly inspect the containers for signs of clogging in the drainage holes or soil compaction. Address any issues promptly to maintain optimal drainage conditions.

Selecting Suitable Growing Media

This is another factor you need to consider carefully. A growing media, also known as a growing medium or substrate, is the material used in containers or beds to support plant growth. This natural soil substitute provides a suitable environment for plant growth.

The growing media should have balanced properties, including aeration, drainage, water retention, and nutrient availability. It anchors the plant roots and supplies them with water, air, and nutrients necessary for healthy development.

Choosing the appropriate growing media ensures optimal conditions for your tree seedlings' growth and overall well-being in a controlled nursery environment. What are the available options, you might ask?

When selecting a suitable growing media for your tree seedlings, you have several available options. Each type of growing media has its advantages and characteristics.

Here are some of the common options to consider:

- **Potting mix:** Potting mix is a well-rounded option that provides a balanced combination of organic and inorganic materials. It offers excellent aeration, drainage, and water retention, making it suitable for varied tree species.

- **Coir fiber:** Coir fiber is an eco-friendly alternative with good water retention and aeration properties.

- **Perlite:** Perlite's lightweight nature enhances aeration and drainage and prevents waterlogged conditions. Many growers usually combine it with other media to create a well-balanced mixture.

- **Peat moss:** Peat moss is a natural material known for its water-holding capacity. It is commonly used in potting mixes to improve water retention, especially for species that need consistent moisture.

- **Pine bark:** Pine bark is cost-effective and provides good drainage and aeration. Its slow decomposition also adds nutrients to the growing media over time.

If you're looking for available and cost-effective growing media options for your tree nursery, consider the following choices:

- **Topsoil:** Local topsoil is often readily available and can serve as a cost-effective growing media. Ensure it is well-draining and free from contaminants.

- **Compost:** You can make compost on-site using organic waste; that makes it an affordable option. It enriches the soil with nutrients and enhances water retention.

- **Pine bark:** Pine bark is a relatively inexpensive option that provides good aeration and drainage. You can purchase it in bulk from local suppliers.

- **Sand:** Sand is a budget-friendly addition to growing media. It improves drainage for plant species that prefer well-drained soil conditions.

- **Peat Substitute:** Peat substitutes, such as coir fiber or coconut coir, are environmentally friendly and affordable alternatives to traditional peat moss.

- **Rice Hulls:** Rice hulls are a cost-effective material that adds aeration to the growing media. They can be a particularly great way to improve drainage.

- **Perlite:** Perlite is a lightweight and affordable volcanic rock that helps with aeration and drainage.

- **Vermiculite:** Vermiculite is another lightweight and cost-effective option that enhances water retention.

- **Sawdust:** If sourced locally, sawdust can be a low-cost material that improves aeration in the growing media.

- **Leaf Mold:** Leaf mold created from composting leaves can be a cost-effective organic addition to the growing media.

You can acquire these growing media options from various sources:

- **Potting mix, coir fiber, perlite, peat moss, pine bark:** You can get these from garden centers, nurseries, home improvement stores, and online retailers specializing in gardening supplies.

- **Topsoil:** Local gardening supply stores, landscaping companies, and agricultural supply centers often sell topsoil.

- **Compost:** You can create compost on-site using organic waste from your nursery or obtain it from composting facilities or garden supply stores.

- **Sand:** Local building supply stores often have different sand types ideal for gardening.

- **Rice hulls, vermiculite, sawdust, leaf mold:** You can source some of these materials locally; for example, you can get rice hulls from rice mills or sawdust from woodworking shops. You can find vermiculite and leaf mold at garden centers and specialty stores.

- **Peat substitute**: Coir fiber or coconut coir is available at garden centers, nurseries, and online gardening retailers.

- **Pine bark:** You can purchase pine bark in bulk from local suppliers, garden centers, and nurseries.

When obtaining growing media, consider the specific needs of the tree species you're cultivating and your local environmental conditions. Read product descriptions, seek recommendations, and ensure the media aligns with your nursery's goals for healthy tree growth.

The aim is to create a growing media that balances good aeration, drainage, and moisture retention, thus supporting the healthy development of your tree seedlings. While cost is crucial, also consider material quality and sustainability to ensure long-term success for your tree nursery.

Other considerations you should factor in include but are not limited to:

- **Nurturing young seedlings:** Once your growing beds or containers are ready, delicately plant the tree seedlings and ensure they settle comfortably in their new homes. Proper placement and gentle handling contribute to their successful acclimation.

- **Monitoring and adjusting**: As your tree seedlings grow, closely monitor their progress. Regularly assess their health and make any necessary adjustments to the growing environment, such as adjusting watering schedules or providing additional support.

- **Nutrient supplementation:** To encourage optimal growth, consider nutrient supplementation. Based on the specific needs of each tree species, you can apply fertilizers or organic amendments to provide essential nutrients for healthy development.

- **Protection and support:** Another thing about young trees is that they may need some extra care and protection. That's why you should protect them from harsh weather conditions or potential threats and give them the best chance to thrive.

- **Timely transplanting:** As your tree seedlings grow and outgrow their current containers or beds, timely transplant them to larger spaces to accommodate their increasing size. This smooth transition ensures their continued development without unnecessary stress.

Irrigation Systems Implementation

Implementing effective irrigation systems is one of the best ways to ensure your tree stock receives the water it needs. Use your nursery's size and layout to select an irrigation method that suits your tree stock's requirements, such as drip irrigation, sprinklers, or overhead misting systems.

Figure 6: Drip irrigation

Figure 7: An example of a sprinkler sprinkling

Figure 8: Example of misting system

Consider drip irrigation because it is an ideal choice for new tree nursery operations; here is why:

While misting and sprinkler systems have their uses in certain situations, they may require more expertise and attention to avoid potential issues like excessive moisture on leaves or uneven water distribution.

Drip irrigation offers beginners a straightforward and effective solution to water management, allowing you to focus on other essential aspects of starting and managing your tree nursery.

Storage Facilities Organization

Finally, another important thing to consider is maintaining a well-run and efficient storage facility within your tree nursery. By having a designated storage space, you'll ensure essential equipment, tools, and nursery supplies are readily available and kept in optimal condition.

To have a well-organized storage facility, implement various storage solutions such as shelves, cabinets, tool racks, and labeled bins. Consider categorizing items based on their purpose and use frequency, and aim to ensure easy access and retrieval. Regularly maintain and clean the storage area to maintain orderliness and prevent clutter accumulation.

When setting up storage facilities within your tree nursery, several key factors and facilities will help you maintain an organized and efficient operation:

- **Tool storage:** Designate an area to store essential gardening tools, equipment, and supplies. Use racks, hooks, and shelves to keep tools organized and easily accessible.

- **Pot and container storage:** Have a space dedicated space to storing pots, containers, and other planting vessels. Use shelves or stackable racks to maximize vertical space and prevent clutter.

- **Growing media storage**: Allocate an area to store bags or bins of various growing media. Arrange them by type and label them for quick identification.

- **Seed and seedling storage:** Create a controlled environment for storing seeds and young seedlings. Use shelves, trays, or containers to prevent damage and maintain viability.

- **Nursery supplies:** Store nursery-specific supplies like labels, tags, ties, and protective materials in labeled bins or drawers.

- **Chemical storage:** If you use fertilizers, pesticides, or other chemicals, store them in a secure and well-ventilated area away from direct sunlight and extreme temperatures.

- **Workspace storage:** Organize workspaces with shelves or cabinets to store notebooks, pens, markers, and any documentation related to your nursery's operations.

- **Seasonal equipment storage:** Plan storage for seasonal equipment like frost covers, shade cloths, and irrigation supplies. Keep them protected during off-seasons.

When designing your storage facilities, consider the following too:

- **Assess your needs:** Evaluate the types and quantities of equipment and supplies you'll be storing. Determine what you need to store indoors and what you can store outdoors with proper protection.

- **Layout planning:** Plan the layout based on accessibility and workflow. Group similar items together and ensure frequently used items are easily reachable.

- **Optimize space:** Maximize space use with vertical storage solutions such as shelves and racks. Use clear bins or containers for easy identification.

- **Labeling:** Label shelves, bins, and containers to indicate the contents; this helps prevent confusion and saves time when locating items.

- **Safety considerations:** When storing chemicals, follow safety guidelines and keep them away from flammable or combustible materials.

- **Climate control:** Consider climate control measures when storing items sensitive to temperature and humidity changes, such as seeds and chemicals.

- **Security:** Ensure the storage area is secure and accessible only to authorized personnel; this prevents theft or unauthorized access.

- **Regular maintenance:** Schedule regular checks and cleanouts to keep the storage area organized and clutter-free.

While setting up a storage facility may involve some initial investment, it doesn't have to cost an arm and a leg. Here are some cost-saving tips:

- **Prioritize essentials:** Begin with storing only the essential tools, equipment, and supplies you'll need to operate effectively. As your nursery grows, you can gradually expand your storage capabilities.

- **DIY solutions:** You can create simple storage solutions using shelves, racks, hooks, and containers. Repurposing existing furniture or using second-hand storage solutions can also be budget-friendly.

- **Start small:** Begin with a basic storage setup and scale up as your nursery business grows and financial resources allow.

- **Optimize space:** Make the most of your available space by using vertical storage solutions and proper organization techniques.

- **Repurpose existing space:** Consider utilizing existing structures like sheds, garages, or covered areas for storage before building or investing in additional facilities.

Remember, a well-organized storage facility doesn't have to be extravagant but is an investment that can significantly enhance your nursery's efficiency, protect your resources, and contribute to your overall business success.

With your nursery carefully prepared, it's time to delve into the heart of your tree nursery operations: cultivation and plant care. The forthcoming chapter will focus on caring for your tree seedlings and assuring a successful green venture. From sowing seeds to providing tender care, we'll discuss every step of cultivating healthy trees. As we move forward, prepare to learn the wonderful art and science of cultivation. Let's get into it.

Chapter 4: Cultivation and Plant Care in Your Tree Nursery

This chapter is one of the most important ones when starting a tree nursery business because we shall focus on the ongoing care and maintenance of the trees in your nursery to ensure optimal health, growth, and overall success.

Proper cultivation and plant care practices ensure your trees develop strong roots, resist diseases, and thrive in their environment. This chapter provides valuable guidance on various aspects of this endeavor, including but not limited to watering techniques, fertilization, pest and disease management, pruning, and overall tree health monitoring.

By mastering the principles outlined in this chapter, you'll know to nurture high-quality trees that attract customers, build your nursery's reputation, and contribute to your business's long-term profitability.

Consistently healthy and well-cared trees are likelier to flourish, increasing their market value and customer appeal. Thus, invest time and effort in understanding and implementing these effective cultivation and plant care practices for a successful tree nursery venture.

Without further delay, let's get into this.

Seedling Propagation

Seedling propagation is one of the fundamental processes in a tree nursery because it lays the foundation that allows you to create a diverse and healthy inventory of trees.

Seedling propagation involves reproducing new trees from seeds, providing a cost-effective and efficient way to expand your tree nursery's inventory. Introducing customers to seedling propagation early on can emphasize the art and science behind growing trees from the beginning stages. This approach highlights your commitment to providing a diverse range of tree species and demonstrates your expertise in cultivating healthy young plants. Successful seedling propagation ensures a consistent supply of young plants, allowing you to offer your customers a wide variety of tree species.

Let's explore some of the key methods used in seedling propagation:

1: Direct Sowing

Direct sowing involves planting seeds directly into the ground or nursery beds. This method is suitable for tree species with seeds that germinate easily without any special treatment. Before sowing, prepare the soil by loosening it and removing any debris.

Consider the requirements of each tree species and plant the seeds at the appropriate depth and spacing. Start by researching the tree species you intend to propagate through direct sowing. Each species may have specific germination and early growth requirements.

If you purchase seeds from a reputable supplier, the seed packets or product information should provide guidelines on the appropriate sowing depth and spacing. This information is usually specific to the tree species you are working with and can serve as a helpful starting point.

You can also consult your local horticulturists, botanists, or experienced nursery operators. They may offer valuable insights into direct sowing practices particularly suited to your regional climate and soil conditions.

For direct sowing:

- Be prepared to adapt your direct sowing practices based on the specific needs of each tree species. Some seeds may require deeper sowing to ensure proper coverage, while others may need shallower depths to receive adequate light for germination.

- Maintain detailed records of your direct sowing activities, including the depth, spacing, sowing date, and specific treatments applied. Regularly monitor seedling progress, noting their growth rate and overall health. This data will empower you to identify successful techniques and needed improvements.

- Don't worry; over time, your experience with direct sowing will improve, and you will become more confident and understand each tree species' requirements. By learning from successes and challenges, you can refine your direct sowing practices and continually improve your nursery's seedling propagation.

Watering and pest protection

Ensure proper watering and protection from pests and adverse weather conditions to support successful germination.

To ensure success with direct sowing:

Watering:

- Keep soil consistently moist but avoid overwatering.

- Water gently with a fine mist or soaker hose to protect the seeds.

- Water in the morning to reduce fungal risk.

- Mulch to conserve moisture and suppress weeds.

Pest protection:

- Regularly monitor for pests.

- Use beneficial insects as natural pest control.

- Use physical barriers to deter animals.

- Opt for pest-resistant tree varieties whenever possible.

Weather protection:

- Use shade cloth in hot and sunny climates.

- Install windbreaks to shield from strong winds.

- Prepare for frost with protective coverings.

- Have a storm contingency plan for adverse weather.

These concise guidelines can help you foster successful germination and create a nurturing environment for healthy seedling growth in your tree nursery.

2: Pre-Treatment Techniques

Some tree species have seeds that benefit from specific pre-treatment techniques to improve germination. These techniques might include soaking seeds in water or treating them with plant growth regulators before sowing. Pre-treatment helps soften seed coats, triggers physiological changes, and creates a more favorable environment for germination.

Common pre-treatments include:

Stratification:

Stratification involves subjecting seeds to cold temperatures to break their dormancy and improve germination rates. Many temperate-climate tree species have seeds that require cold stratification to simulate real winter conditions; without breaking this hibernation, they will not begin to grow. These indigenous seeds spend the winter months in the ground in the wild, where frost and other natural exposure to the environment soften their coats. This damp, cold stage activates the seed embryo, and as it grows and expands, it finally penetrates the weakened seed coat in quest of sunlight, warmth, and nourishment.

To stratify seeds, place them in a moist medium, such as peat moss or sand, and store them in a refrigerator or cold room for the recommended duration. The recommended duration for stratifying seeds can vary depending on the tree species. Different tree species have different dormancy periods, and the time they need to stratify can range from a few weeks to several months.

To determine the specific or recommended duration to stratify a particular tree species, refer to reliable sources of information. Some common resources for this information include seed supplier guidelines, horticultural books, agricultural extension offices, and forestry departments. These sources often provide detailed instructions on the optimal stratification period for various tree species.

When planning your seedling propagation, make sure to identify the tree species you intend to stratify and research the recommended duration for their dormancy treatment. Adhering to the appropriate stratification period can enhance germination rates and increase the overall success of your tree nursery. After stratification, use the appropriate planting guidelines to sow the seeds.

Scarification:

Certain tree seeds have hard seed coats that inhibit water absorption and germination.

Scarification involves mechanically breaking or thinning the hard seed coat to enhance water penetration and encourage germination. By employing scarification techniques, you can significantly increase the chances of successful seedling establishment, leading to a more diverse and vibrant inventory in your tree nursery.

Scarification makes it possible to propagate tree species that would otherwise struggle to germinate, expanding the range of trees you can offer your customers. Additionally, scarification allows for more efficient use of seeds by reducing waste and increasing the yield of viable seedlings.

Incorporating scarification into your seedling propagation practices demonstrates your expertise as a nursery operator because it shows you understand specific tree species' unique germination requirements. By mastering this technique, you can produce healthy and robust seedlings, setting the stage for the success and reputation of your tree nursery in the horticulture market.

You can use different scarification methods, including abrasion with sandpaper, nicking with a knife, or soaking seeds in hot water. The appropriate scarification method depends on the tree species and the specific characteristics of the seed coat. Different tree seeds have different levels of dormancy and respond differently to scarification.

To perform scarification effectively and meet the standards:

- **Research the tree species:** Understand the specific scarification requirements of each tree species. Consult reputable sources or refer to seed supplier guidelines for the best scarification method.

- **Safety and precision:** When using a knife or a sharp tool to nick seeds, exercise caution to avoid damaging the seed embryo. The goal is to thin the seed coat without harming the vital part inside.

- **Trial and observation**: Conduct small-scale scarification trials with different seeds and methods to observe their responses. Monitor germination rates and seedling vigor to determine the most successful scarification approach.

- **Consistency and documentation:** Record the scarification methods used, duration, and other observations for future reference. Consistency in your scarification practices will help maintain high-quality seedling production.

- **Timing:** Perform scarification appropriately at the right time, considering the species' natural germination cues and the desired planting schedule. Consider your nursery's production schedule and scarify seeds with enough time for germination to align with your planting plans. By timing scarification correctly, you can optimize germination potential and ensure successful seedling growth in your tree nursery.

Here are some tips to help you ensure successful seed propagation:

- You can ensure successful seedling propagation by understanding the specific requirements of each tree species and tailor your approach accordingly. Keep detailed records of your seed sowing activities by recording the sowing date, the seed treatment used, and the germination rates. Monitoring and adjusting your propagation methods based on the results will enhance your seedling production.

- As your seedling propagation experience increases, you will become adept at recognizing the unique needs of different tree species. This knowledge will empower you to produce a diverse inventory of healthy and robust seedlings that appeal to your customers.

- Remember that maintaining a carefully managed seedling propagation program is a continuous process. Regularly review your techniques and seek opportunities to refine your methods based on feedback from your nursery's success. With dedication, attention to detail, and a passion for cultivating trees, your nursery will thrive with a remarkable assortment of young, vibrant seedlings ready to grow into majestic trees.

Fertilization Strategies

Another strategy you must develop is a fertilization plan to provide essential nutrients for your plants' healthy growth. Understand the nutrient requirements of various tree species and use fertilizers that promote optimal root development and foliage health.

The first step to developing a fertilization plan for your tree nursery is understanding the nutrient requirements of different tree species. Conduct soil tests to determine current nutrient levels and Ph. Use the results to choose fertilizers that address specific deficiencies.

Use balanced fertilizers with essential nutrients like nitrogen, phosphorus, potassium, and other micronutrients. For young seedlings, opt for fertilizers that promote root development. Mature trees may benefit from slow-release or organic fertilizers. Regularly monitor plant health and use what you learn to adjust the fertilization plan to ensure optimal growth and vitality. Consult horticultural experts for further guidance on choosing the right fertilizer for your tree nursery.

Pest and Disease Management

Identify common pests and diseases that can affect tree seedlings in your region, then implement integrated pest management (IPM) strategies, combining biological, cultural, and chemical methods to minimize pest and disease impact.

- **Biological control:** Use natural predators, parasites, or pathogens to control pest populations. For example, use beneficial insects that feed on pests to maintain a balanced ecosystem.

- **Cultural practices:** Implement cultural practices that reduce the likelihood of pest and disease outbreaks. This includes techniques like planting resistant tree species, proper spacing to avoid overcrowding, and practicing good hygiene to prevent disease spread.

- **Chemical control:** Although you should try to minimize chemical interventions, it may sometimes be necessary. When using pesticides or fungicides, opt for environmentally friendly options and apply them judiciously according to label instructions.

Regarding pest and disease control, remember to:

- Regularly inspect your tree seedlings for signs of pests or diseases. Early detection will make it easier to take swift action and prevent issues from escalating.

- Diagnostics: If you encounter an unfamiliar problem, seek guidance from local agricultural extension offices, nurseries, or experts who can help you accurately diagnose the issue and determine appropriate remedies.

- Record keeping: Maintain meticulously detailed records of pest and disease occurrences, treatments applied, and their effectiveness. These records will help you track trends over time and refine your management strategies.

- Environmental considerations: Be mindful of your nursery's ecosystem. Promote biodiversity by planting companion plants that attract beneficial insects, create habitats for natural predators, and foster overall ecosystem health.

- Training and education: Keep yourself and your staff informed about the latest pest and disease management techniques through workshops, seminars, and industry resources.

Monitoring and Record Keeping:

Maintain thorough records of your nursery operations, including planting dates, watering schedules, fertilization, pest control measures, and plant performance. Regular monitoring and record-keeping enable you to track progress and make informed decisions.

Here's how you can effectively manage these records:

Organized System

Set up a systematic record-keeping approach. Use notebooks, digital spreadsheets, or specialized nursery management software to organize your records.

Several specialized nursery management software options can help you organize your records efficiently. Some popular choices include:

- **Nursery Notes:** Nursery Notes is designed specifically for nurseries; it helps manage plant inventory, sales, and tasks. It offers features like barcode scanning, photo attachments, and customizable reports.

- **Nursery Management Software by DynaSCAPE:** This software offers tools for plant inventory management, sales tracking, and estimating. It's suitable for both small and large nurseries.

- **Floracraft:** Floracraft provides solutions for inventory management, sales orders, and customer communication. It also includes a mobile app for on-the-go access.

- **NurseryBiz:** NurseryBiz offers numerous features like inventory tracking, plant labeling, and customer management. Its unique design seeks to streamline nursery operations.

- **Nursery Index:** Nursery Index is an inventory management system with features like barcode scanning, sales tracking, and reporting capabilities.

- **Tree Tracker:** Tree Tracker focuses on inventory management and provides tools to track tree information, maintenance, and sales.

- **Arborgold:** Arborgold is a comprehensive business management software for tree care and landscaping companies. It offers features like estimating, scheduling, invoicing, and more.

You can easily find these specialized nursery management software options by searching for the tools on popular search engines or visiting their official websites. This will give you access to more information, features, and possibly trial versions that help you settle on the best software for your nursery management needs.

When choosing nursery management software, consider your specific needs, the size of your nursery, and the features that align with your business goals. It's a good idea to explore the websites of these software providers to learn more about their offerings and how they can benefit your tree nursery.

Planting dates

Document when you plant each batch of tree seedlings. Note down the species, quantity, and any relevant details.

Watering schedules

Create a watering schedule that outlines when and how much water each batch of seedlings receives and ensures consistent and appropriate irrigation.

Fertilization

Record the type of fertilizer used, application rates, and fertilization dates to ensure ideal nutrient management.

Pest control measures

Detail the types of pests or diseases encountered, the control methods used (biological, cultural, or chemical), and their effectiveness.

Plant performance

Regularly assess your nursery's and seedlings' growth, health, and overall performance. Note any signs of stress, disease, or other issues.

Weather conditions

Track weather conditions, including temperature, humidity, and rainfall. These factors can influence plant health.

Observations

Write down any observations, changes, or anomalies you notice in your nursery. This information can provide insights into patterns and trends.

Photographic documentation

Supplement your records with photographs. Visual documentation can help you compare growth over time and better understand changes.

Regular Updates

Dedicate specific times for record updates. This might be daily, weekly, or monthly, depending on the activity level in your nursery.

Consistency

Ensure that whoever is responsible for nursery operations follows the record-keeping process consistently to maintain accuracy and continuity.

Review and analysis

Regularly review your records to identify patterns, successes, and areas for improvement. Use this analysis to make informed decisions about adjustments to your cultivation practices.

Backup

For digitally maintained data records, regularly back up your records to prevent data loss.

Training

Train your staff on the importance of record keeping and the specific details they need to document.

Maintaining thorough and organized records helps create a valuable resource that guides your nursery's growth, helps troubleshoot problems, and facilitates long-term planning.

Environmental Sustainability

Promote sustainability in your nursery by conserving resources, recycling, and minimizing waste. Consider incorporating organic and environmentally friendly practices in your cultivation methods.

Here are some practices to consider:

- **Organic fertilizers:** Use organic fertilizers such as compost, manure, and composted plant materials. These materials release nutrients slowly and improve soil structure over time.

- **Mulching**: Apply organic mulch around your seedlings. Mulch promotes soil moisture retention and weed suppression and adds nutrients to the soil as it breaks down.

- **Companion planting:** Introduce companion plants that support the growth of your tree seedlings. Some plants can deter pests, improve soil health, and create a habitat that promotes beneficial insects.

- **Biological pest control:** Encourage natural predators and beneficial insects to control pests. Ladybugs, lacewings, and predatory nematodes can help keep pest populations in check.

- **Crop rotation:** If you have multiple growing beds, practice crop rotation to prevent the buildup of pests and diseases in the soil.

- **Integrated Pest Management (IPM):** Implement IPM strategies focused on prevention, monitoring, and using the least harmful interventions first.

- **Natural pest repellents:** Use natural pest repellents like soap solutions, garlic spray or neem oil to deter pests without harming the environment.

- **Organic soil amendments:** Incorporate organic soil amendments like kelp meal, bone meal, and rock phosphate to enrich the soil with essential nutrients.

- **Rainwater harvesting:** Collect rainwater for irrigation purposes. Rainwater is chlorine-free and better for your plants and the environment.

- **Beneficial microbes:** Introduce beneficial microbes to the soil through compost or microbial inoculants. These microbes improve soil structure and promote nutrient uptake.

- **Cover crops:** Plant cover crops during off-seasons to improve soil health, prevent erosion, and provide habitat for beneficial organisms.

- **Minimal synthetic chemicals:** Reduce your use of synthetic chemicals, such as pesticides and herbicides, and opt for organic alternatives.

- **Natural pest traps:** Set up natural pest traps, like sticky traps, to monitor and control pest populations.

- **Native plant selection:** Choose native plant species for your growing beds. Native plants have adapted to local conditions and require fewer inputs.

- **Responsible watering:** Use efficient irrigation methods like drip irrigation to minimize water waste.

- **Beneficial habitats:** Create habitats for beneficial insects, birds, and other wildlife that contribute to a balanced ecosystem.

By implementing these practices, you contribute to a healthier ecosystem, reduce environmental impact, and produce tree seedlings equipped to thrive in natural landscapes.

Quality Control and Plant Inspection

Implement a stringent quality control process that ensures your nursery stock meets the highest standards. A rigorous quality control process helps maintain excellent nursery stock.

Consider incorporating the following practices:

- **Regular inspections:** Conduct thorough visual inspections of your tree seedlings at different growth stages. Check for signs of diseases, pests, deformities, or nutrient deficiencies.

- **Documentation:** Keep detailed records of each tree species, including planting dates, growth milestones, and any interventions such as fertilization or pest control.

- **Health certification:** Obtain health certificates or phytosanitary certificates when required by regulations. This ensures your nursery stock is free from pests and diseases that could spread to other locations.

- **Sampling:** Regularly take a subset of your nursery stock for in-depth assessment. This helps you identify potential issues before they affect your entire inventory.

- **Testing:** Use laboratory tests to measure soil health, disease presence, and nutrient content to identify problems early.

- **Isolation:** Isolate any new plants before introducing them to your main nursery area to prevent and limit the spread of potential diseases or pests.

- **Quarantine:** Establish a quarantine area for new plant material; this will allow you to observe plants before integrating them into your main nursery stock.

- **Certified seeds:** Go out of your way to source seeds from reputable suppliers who provide certified and disease-free seeds to ensure the quality of your initial planting material.

- **Employee training:** Train your staff to recognize signs of poor health, diseases, and pests, and empower them to take appropriate measures.

- **Customer feedback:** Encourage feedback from customers who purchase your tree stock. This can help identify any issues your inspection process may have missed.

By implementing these quality control measures, you'll safeguard the health and vitality of your nursery stock, build customer trust, and enhance the reputation of your tree nursery.

In conclusion:

Cultivation and plant care are fundamental aspects of running a successful tree nursery. Mastering seedling propagation, transplanting, irrigation, fertilization, pest and disease management, and adopting sustainable practices ensures you can produce healthy and vibrant trees.

Consistently providing top-quality trees will delight your customers and establish your nursery's reputation as a reliable and reputable source of healthy trees.

Now that we have learned the essential aspects of cultivating and caring for your tree nursery, it's time to shift our focus to another critical aspect of the business: marketing and sales.

Chapter 5: Sales and Marketing Strategies for Your Tree Nursery

This chapter will explore essential marketing and sales strategies to promote and grow your tree nursery business.

Effective marketing techniques will help you reach your target audience, create brand awareness, and generate sales leads. These strategies raise awareness about your unique tree varieties, foster customer interest, and drive sales. By implementing these strategies, you can elevate your nursery's visibility in the market and build a strong customer base.

Where do you start?

Identify Your Target

Identifying your target market is like finding your tribe in the vast forest of potential customers. Take the time to know them inside out! Figure out who they are, what they want, and the trees, landscaping, and gardening concerns that keep them up at night.

Start by defining your ideal customer profile – think about their age, location (where they stay), interests, and lifestyle (do they enjoy spending time in their gardens or value the beauty of nature in their outdoor spaces?). Then, dig deeper into their preferences and needs regarding their choice of trees for their yards or gardens. What types of trees do they love? Are they looking for shade, privacy, or ornamental trees?

Understanding their pain points is also important: What challenges are they facing when choosing their ideal trees? What challenges do they face when it comes to landscaping or garden design? Are they worried about maintenance, budget constraints, or finding the right trees for their specific climate?

Once you have all this valuable information, tailor your marketing messages to them. Speak their language, address their concerns, and offer solutions that make their dreams come true. Whether you do it through your website, social media, or ads, ensure your message resonates with your target audience and shows them you're the tree expert that's been missing in their lives. With this level of personalization, you'll package your offerings in ways that win their hearts!

Develop a Brand Identity

Why is this important?

Developing a brand identity for your nursery is like crafting a distinctive personality that sets it apart from all other brands. Just like every person has a unique fingerprint, your nursery's brand identity should be one-of-a-kind and easily recognizable.

- Start by creating a cool logo that embodies the essence of your nursery's mission and values. The logo should be simple, memorable, and visually appealing, reflecting the beauty of trees and nature. A well-designed logo will become the symbol of your nursery, appearing on signage, packaging, and promotional materials.

- Next, create a catchy tagline that captures your nursery's unique selling proposition. The tagline should be short, snappy, and memorable enough to convey what makes your nursery special. It should evoke emotions and make a lasting impression on potential customers. i.e., "Where Nature's Beauty Takes Root."

- Visuals play a crucial role in brand identity. Use captivating visuals and imagery that align with your nursery's brand personality and target audience. Use consistent and eye-catching visuals on your website, social media posts, and marketing materials to leave a lasting impression on customers.

- Your brand identity should reflect the values and promises you make to your customers. Are you all about providing top-quality tree seedlings and exceptional customer service? Let your brand identity convey that message.

When done right, a strong brand identity becomes a magnet that draws customers in with its unique appeal and fosters a sense of trust and loyalty. Like a superhero cape, your nursery's brand identity will help your business soar to new heights, stand out in the crowded market, and leave a memorable mark on your customers' hearts.

Website and Online Presence

Here, focus your marketing energy on the following core aspects—but please also tailor your marketing approach to your brand and target audience:

Online Marketing

Take a multi-faceted online marketing approach that encompasses the following:

Website marketing for your tree nursery

A well-optimized website is a cornerstone of your tree nursery's online presence. It acts as a virtual storefront that can attract, engage, and convert potential customers.

Your website is often the first interaction customers have with your nursery. A user-friendly and visually appealing design establishes a positive first impression and increases the likelihood of further exploration. In addition, it helps your customers access comprehensive details about your tree species, care instructions, and benefits at their convenience. This transparency builds trust and helps customers make well-informed purchasing decisions.

But how do you make the most out of it?

Here's how to make the most of your website:

- **User-Friendly Design:** Please create an intuitive, visually appealing, easy-to-navigate website. Ensure it is responsive and mobile-friendly for seamless browsing across devices.

- **Product Showcase:** Display high-quality images and detailed information about your tree species. Include descriptions, care tips, and benefits to help customers make informed decisions.

- **Content Creation:** Regularly update your website with informative blog posts, articles, and guides about tree care, landscaping ideas, and gardening tips; this establishes your expertise and engages visitors.

- **E-Commerce Integration:** If you plan to sell trees online, integrate an efficient e-commerce system for easy ordering, payment, and delivery processes.

- **Customer Testimonials:** Showcase positive feedback from satisfied customers to build trust and credibility.

- **Contact Information:** Provide clear contact details and a user-friendly contact form for inquiries and assistance.

To make this possible, you can choose a website platform or hire a web developer experienced in creating user-friendly designs. The cost of choosing a website platform or hiring a web developer experienced in creating user-friendly designs can vary based on several factors.

Website platforms like WordPress, which offer user-friendly templates, can be cost-effective. However, if you require advanced features or customization, you might need to invest in premium themes or plugins, which can add to the cost.

Hiring a web developer can attract varying costs depending on the developer's expertise, location, and the complexity of the website. Rates can range from affordable to higher-end, depending on your requirements. It's essential to consider your budget and the specific needs of your tree nursery's website.

While there might be some initial investment, a well-designed and functional website can provide long-term benefits, attracting customers and boosting your online presence. Please thoroughly research your options and weigh the potential returns on investment before settling on an option.

Social media marketing for your tree nursery

A social media platform is another powerful engagement tool that can showcase your tree nursery. It allows you to interact with your customers, answer their queries, and build a loyal community. Engaging content fosters meaningful connections and increases customer trust.

Here's how to leverage social media effectively:

- **Visual Content:** Use high-quality images and videos of your trees, nursery facilities, and staff. Visual content is highly engaging and shareable on social media. The best way to achieve this is to hire a professional photographer to capture appealing shots of your trees, nursery environment, and staff, but you can also DIY it if you have a good camera. Professional equipment and expertise ensure top-notch visuals. Alternatively, you can use a modern smartphone with a good camera capable of taking high-quality photos. After capturing great images, use photo editing apps or software to enhance and adjust brightness, contrast, and colors to make your visuals pop.

- **Platform Selection:** Choose platforms used by your target audience. Facebook, Instagram, Pinterest, and Twitter can be particularly effective for nurseries.

- **Content Variety:** Share posts about tree care, gardening tips, nursery updates, customer stories, and behind-the-scenes content. Keep a balance between promotional and informative posts.

- **Consistency:** Regular posting and engaging with your audience are the secret to maintaining a strong social media presence. Use scheduling tools to ensure consistent updates.

- **Engagement:** Respond promptly to comments, messages, and inquiries. Encourage user-generated content by featuring customer photos with your trees.

- **Hashtags:** Use relevant hashtags to increase post discoverability. Research popular gardening and landscaping hashtags to ensure you can effectively use them after identifying the commonly used ones within the gardening and landscaping communities. Incorporate a mix of broader and more specific hashtags to maximize reach. Remember that each social media platform may have unique hashtag usage trends and etiquette, so tailor your approach accordingly. Using hashtags strategically allows you to expand your social media reach, engage with a wider audience, and ultimately attract potential customers to your tree nursery.

- **Paid Advertising:** Consider using paid advertising on social media to target specific demographics and expand your reach. Paid social media advertising involves creating and running targeted ads to reach a specific audience on platforms like Facebook,

Instagram, Twitter, and more. This method allows you to tailor your ads to reach users based on demographic factors like age, location, interests, and behavior.

When you use paid advertising, you can select the demographics of the audience you want to target, ensuring your ads appear to potential customers who are more likely to be interested in your tree nursery products. These ads can appear in users' feeds, stories, or as sponsored content.

Paid social media advertising offers advantages such as increased visibility, precise audience targeting, and the ability to track and measure the performance of your campaigns. By investing in paid advertising, you can effectively promote your nursery to a broader audience, attract new customers, and boost your online presence.

Successful Examples:

- **Website Marketing:** Check out The Tree Center's website for a comprehensive showcase of tree species, care guides, and an efficient e-commerce system.[10]

- **Social Media Marketing:** Davey Tree's Instagram showcases their tree services and tree care tips and effectively engages with followers.[11]

Search Engine Optimization (SEO)

Have you heard of SEO (Search Engine Optimization)? Search Engine Optimization (SEO) involves on and off-page endeavors to increase your website's visibility on search engines like Google. It entails optimizing various aspects of a website to rank better in search results, resulting in more organic (non-paid) visitors to the site.

Although it closely relates to website marketing, it encompasses techniques such as keyword optimization, improving website structure, meta tags, image optimization, and creating valuable content to attract both search engines and users.

---

[10] https://www.thetreecenter.com/
[11] https://www.instagram.com/daveytree/

SEO is important because when potential customers search for keywords related to your tree nursery, a well-optimized website will appear higher in the search results, thereby increasing the chances of attracting organic traffic.

Here's how you can implement effective SEO strategies:

- **Keyword research:** Identify relevant keywords that potential customers might use to search for tree nurseries or related topics. Use tools like Google Keyword Planner[12] to find high-volume and relevant keywords.

- **On-Page Optimization:** Optimize each website page by including target keywords in page titles, headings, meta descriptions, and throughout the content. However, prioritize natural, reader-friendly content over keyword stuffing.

- **Quality content:** Always ensure you create content that is of high-quality, informative, and engaging for the audience. It should also address the target audience's needs and questions. This attracts visitors and encourages them to spend more time on your site, which is a positive ranking signal for search engines.

---

[12] https://ads.google.com/home/tools/keyword-planner/

- **Website structure and navigation:** Ensure your website is user-friendly and easy to navigate. Organize your content logically with clear headings and subheadings. A well-structured website enhances user experience and helps search engines understand your content.

- **Mobile responsiveness:** With more users accessing the web from mobile devices, you would do well to create a mobile-responsive website. Google considers mobile-friendliness when ranking websites.[13]

- **Page Loading Speed:** Improve your website's loading speed as it affects user experience and search engine rankings. Compress images, minimize plugins, and use a reliable hosting provider.

- **Link Building**: Acquire high-quality backlinks from reputable websites in the same or related industry. These backlinks tell search engines that your website is credible and authoritative.

---

[13] https://www.searchenginejournal.com/ranking-factors/mobile-friendliness/

- **Local SEO:** Optimize your website for local searches if you have a physical nursery. Claim your Google My Business[14] listing, ensure accurate contact details, and encourage customer reviews.

- **Analytics and Monitoring:** Use tools like Google Analytics to monitor website traffic, user behavior, and keyword performance. This data helps you refine your SEO strategies over time.

- **Stay Updated:** SEO trends and algorithms change over time. Stay informed about the latest updates and adjust your strategies accordingly.

Content Marketing

Content marketing involves creating and distributing informative and valuable content to your target audience to build brand awareness, establish brand authority, and engage potential customers.

---

[14] https://www.google.com/business/

In the context of a tree nursery business, content marketing can include various forms of content:

- **Blog Posts and Articles:** You can regularly publish blog posts and articles on your website that provide helpful information about tree care, landscaping, gardening tips, and other relevant topics. These pieces should be well-researched, informative, and full of practical insights.

- **Guides and eBooks:** Create in-depth guides or ebooks that cover specific topics in detail. For example, you could offer a guide on selecting the right trees for different landscapes or a comprehensive ebook about proper tree care.

- **Videos:** Creating video content about tree care, planting techniques, nursery tours, and more can engage viewers and help showcase your expertise. Here are some creative ways to create and use video content:

  - Create step-by-step tutorials demonstrating proper tree planting techniques, pruning methods, or caring for specific tree species. These videos can provide valuable guidance to your audience and position your nursery as an authority on tree care.

o You can also take viewers on a virtual tour of your tree nursery. Showcase your facilities, the variety of trees you offer, and the care practices you implement. This personal and behind-the-scenes glimpse can build a stronger connection with potential customers.

o Feature videos of satisfied customers sharing their experiences with your nursery. Authentic testimonials can build trust and credibility and assure potential buyers of the quality of your products and services.

o You can develop a series of educational videos covering different topics related to trees, landscaping, and gardening. This could include seasonal planting tips, disease prevention methods, or selecting the right trees for specific environments.

o Host live or recorded Q&A sessions where you answer common questions from your audience about tree care, maintenance, and other relevant topics. This interactive approach can foster engagement and build a sense of community.

o Share real-life examples of successful tree-planting projects or challenging scenarios you've tackled. Discuss the strategies you used and the outcomes achieved, providing valuable insights for viewers.

o Showcase the features and benefits of specific tree species or tree-related products in your nursery. Highlight what makes them unique and how they can enhance a customer's landscaping.

o **TIP:** When creating these videos for content marketing, aim to maintain a clear-cut balance between informative and engaging content. Keep your videos visually appealing, concise, and relevant to your target audience. Incorporate storytelling elements, use clear visuals and graphics, and ensure high production quality to leave a positive and lasting impression. Sharing your videos across various platforms, including social media and your website, can help maximize their reach and impact.

- **Podcasts:** If you're comfortable with audio content, hosting a podcast where you discuss tree-related topics and gardening trends and answer listener questions can be a unique way to connect with your audience.

- **Social Media Posts:** Craft informative and engaging posts for your social media platforms that provide quick tips, facts, or insights about tree care and gardening.

Content marketing is a great way to offer valuable information that addresses the needs and interests of your target audience. Establishing your nursery as a reliable source of expertise builds trust and credibility and attracts potential customers seeking reliable information and solutions. Over time, content marketing can be one of the best ways to build a loyal customer base and increase your overall reach and online presence.

Online listings and directories

List your tree nursery in local business directories, gardening websites, and online marketplace platforms. Ensure your business information is accurate and up-to-date to improve your online presence. Promptly update any changes in business hours, services, or contact details to avoid inconveniencing potential customers.

Potential customers rely on these listings to find essential details such as your nursery's name, address, contact information, and website link. Having accurate information builds trust and ensures interested customers can easily get in touch with you.

Optimize your listings with captivating images and engaging descriptions. A visually appealing profile with compelling content entices users to explore more about your nursery.

As customers browse these directories and platforms, they'll find your nursery listed alongside competitors. Stand out from the competition by providing unique selling points and highlighting what makes your nursery special. Showcasing positive customer reviews and ratings adds credibility and helps build trust.

Offline marketing

When it comes to offline marketing, you have options like:

Promotional events and open houses

Promotional events and open houses can be powerful marketing tools for your tree nursery business because they offer a unique opportunity to engage directly with your target audience and create a memorable experience.

Here's how to make the most of these events:

- **Special discounts and offers**: Entice visitors with exclusive discounts or promotions on select tree species, products, or services during the event. Limited-time offers can create a sense of urgency and encourage attendees to purchase.

- **Demonstrations and workshops:** Plan interactive demonstrations and workshops that showcase your expertise in tree care, planting techniques, and landscaping. These sessions provide valuable educational content and allow attendees to learn firsthand from your knowledgeable staff.

- **Nursery tours**: Conduct guided tours of your nursery, highlighting different tree varieties, growth stages, and care practices. This gives attendees a comprehensive view of your offerings and allows them to see the quality of your products.

- **Hands-on activities:** Engage attendees with hands-on activities such as tree planting workshops, potting demonstrations, or pruning tutorials. This interactive approach fosters a deeper connection with your tree nursery and encourages participation.

- **Expert consultations:** Offer one-on-one consultations with your knowledgeable staff. Attendees can discuss their specific landscaping needs, ask questions, and receive personalized recommendations for the best tree species for their environment.

- **Networking opportunities:** Promotional events provide a platform for attendees to connect with like-minded individuals, including fellow gardening enthusiasts, landscapers, and local community members.

- **Brand exposure:** Promote your nursery's brand identity through signage, banners, and promotional materials displayed throughout the event. This reinforces brand recognition and establishes a professional image.

- **Engagement with attendees:** Encourage attendees to share their experiences on social media by creating event-specific hashtags. This extends your event's reach and enhances online engagement.

- **Collect customer feedback**: Take advantage of the event to gather valuable feedback from attendees. This can provide insights into their preferences, needs, and expectations, helping you tailor your offerings accordingly.

- **Follow-up opportunities:** Follow up and stay connected with attendees through email newsletters or social media updates after the event. Tell them about upcoming promotions, new arrivals, and relevant educational content.

Promotional events and open houses attract potential customers and can help you build stronger relationships with your existing clientele. They provide a face-to-face platform to showcase your expertise, demonstrate the value of your products, and create a sense of community around your nursery.

By offering a mix of educational content, interactive activities, and special offers, you can create a memorable experience that encourages attendees to become loyal customers.

Referral programs and customer loyalty

Referral programs and customer loyalty initiatives are powerful strategies that can significantly boost your tree nursery business by leveraging the goodwill of your existing customers:

1. Referral Programs

Encourage your current customers to refer friends, family, and acquaintances to your nursery. Incentivize referrals by offering discounts, free products, or other rewards for successful referrals. This brings in new customers and showcases the trust your existing customers have in your products and services.

It's essential to make the referral process easy and accessible; you can do that through dedicated referral codes, personalized links, or simple word-of-mouth.

- **Referral program implementation:** Create clear guidelines for your referral program by outlining how customers can refer others and what rewards they will receive. Communicate the program through various channels, including your website, social media, and in-store materials. Ensure that customers understand the value they'll gain by referring others, making it a win-win situation.

2. Customer loyalty programs

Reward your repeat customers for their continued support and purchases.

A customer loyalty program can take various forms, such as a points-based system where customers earn points for every purchase and redeem them later for discounts or free items. You can also offer exclusive deals, early access to new products, or special events for loyal customers. These programs enhance customer retention and encourage them to choose your nursery over competitors.

- **Customer loyalty program design:** Tailor your loyalty program to suit your customers' preferences. Determine how customers will earn and redeem points and showcase the benefits they'll receive. Use data and insights from your sales history to identify your most loyal customers and personalize rewards to suit their preferences.

  o **Tracking and management:** Use technology to track referrals and loyalty points accurately. Implement a system that records referrals and awards points automatically. This streamlines the process and makes customers' efforts feel recognized.

  o **Promotion and communication:** Regularly promote your referral and loyalty programs across various touchpoints. Use email marketing, social media posts, and even physical signage at your nursery to remind customers of the benefits of participating.

  o **Monitor and adjust:** Continuously assess the effectiveness of your programs. Monitor the number of referrals, the engagement of your loyal customers, and the impact on sales.

Gather customer feedback to understand what works well and what could be improved.

Referral programs and customer loyalty initiatives can foster a sense of community around your tree nursery while driving customer engagement and growth. By valuing and rewarding your customers' loyalty and advocacy, you can create a positive cycle where satisfied customers become your best promoters, helping your nursery thrive.

Visual merchandising and displays

Create visually appealing nursery displays highlighting various tree species and their features. Well-organized and attractive displays can inspire customers and make the tree selection process enjoyable.

Seasonal marketing campaigns

Tailor your marketing efforts to seasonal trends and events— plan campaigns for peak planting seasons, holidays, or special occasions to capitalize on increased customer interest.

Holidays are like magical occasions that bring people together. Plan themed campaigns that evoke festive feelings and showcase tree species customers consider perfect for holiday decorations or gifting. Offer special discounts or gift packages to entice customers seeking unique and meaningful presents.

Special occasions like Arbor Day or Earth Day are our way to celebrate nature's wonders. Design campaigns that emphasize the significance of these events and encourage customers to join in the green movement. Organize tree-planting events or workshops to engage the community and promote environmental awareness.

Use social media platforms to sow the seeds of anticipation for your seasonal campaigns. Tease upcoming promotions and events to generate excitement and curiosity among your audience. Measuring the success of your seasonal campaigns is a lot like evaluating your plants' growth. Analyze data, such as website traffic, sales, and customer feedback, to identify what worked well and what you can improve in future campaigns.

In addition to the strategies mentioned, you can also consider the following to enhance your marketing and sales efforts for your tree nursery:

- **Customer engagement and personalization:** Use customer feedback and insights to personalize your interactions and offerings. Engage with customers through email marketing, newsletters, or loyalty programs to provide access to exclusive offers and relevant information.

- **Collaborations and cross-promotions:** Partner with local businesses or organizations for joint promotions or cross-promotions. This can help increase your reach and attract customers from different niches.

- **Customer testimonials and case studies:** Highlight success stories and positive experiences from satisfied customers. Showcase these testimonials and case studies on your website and social media to build trust and credibility.

- **Environmental sustainability efforts:** Emphasize your nursery's commitment to eco-friendly practices and sustainable growing methods. Many customers value environmentally responsible businesses—this can be a compelling selling point.

- **Seasonal contests and giveaways:** Organize fun and interactive contests or giveaways during peak seasons to create excitement and engagement. Offer tree-related prizes or special discounts to winners or participants.

- **Market research and competitive analysis:** Continuously research your competitors and monitor the market. Understanding market trends and your competitors' strategies can help you make informed decisions and stay ahead in the industry.

By incorporating these additional elements into your marketing and sales strategies, you can further strengthen your tree nursery's position in the market and attract a loyal customer base. The key is continuously adapting and evolving your approaches to meet changing target audience needs and preferences.

Having explored effective sales and marketing strategies, let's now focus on pricing your plants for success and learn how to set prices that attract customers while ensuring profitability for your thriving tree nursery. Are you ready to unlock pricing secrets and achieve sustainability in your business? If YES, Let's dive in!

Chapter 6: Pricing Your Plants for Success

Having concluded our exploration of effective sales and marketing strategies, we can now turn our attention to an equally vital aspect of your tree nursery business: pricing your plants for success. Just as a well-crafted marketing plan can attract customers, a carefully devised pricing strategy can determine your nursery's profitability and sustainability.

This chapter will delve into the art of pricing your plants and equip you with essential knowledge and practical strategies to help you set prices that strike the perfect balance between customer satisfaction and business growth. By understanding your costs, researching the market, and adopting suitable pricing approaches, you will have all the knowledge you need to navigate pricing complexities in the tree nursery industry.

Let's now embark on this journey together and unlock the power of pricing in establishing a thriving and sustainable tree nursery business, especially because finding the sweet spot between affordability and profitability can be a bit like walking on a tightrope—fear not; we've got you covered!

First things first, know your costs inside out.

Calculate everything

From the initial investment in seeds or seedlings to nurturing them into beautiful trees, factor in labor, materials, and any overhead expenses, including wages for employees involved in planting, watering, pruning, and other tasks.

Beyond direct costs, consider overhead expenses like utilities, rent, or mortgage for your nursery space, marketing expenses, and administrative costs.

A comprehensive understanding of your costs can help you establish a solid pricing strategy that ensures your plant prices cover all your expenses while generating a reasonable profit margin.

However, remember that pricing too high can drive potential customers away, while pricing too low may lead to insufficient revenue to sustain your business.

Striking the right balance will help keep your nursery financially healthy and position your plants competitively in the market.

Consider your target audience

Next, consider your target market. Are your primary customers bargain hunters looking for budget-friendly options, or do they value premium quality and uniqueness?

Understanding your customers' preferences will guide your pricing decisions. How so?

For budget-conscious customers, offering competitive pricing and cost-effective options may be the key to attracting them to your nursery. Consider offering a range of plant sizes and varieties at different price points, ensuring there's something for everyone.

On the other hand, if your target audience values premium quality and unique plant species, you can justify higher prices by emphasizing your plants' exceptional features, rareness, and benefits. Showcase each plant's value to the customer's landscape or garden, such as increased curb appeal or unique visual appeal.

Additionally, you might offer specialty services, such as personalized plant care advice or landscape design consultations, to cater to customers seeking an elevated experience and willing to pay a premium.

Do a thorough competitive analysis

Another thing you have to look into is competitive analysis, which is your best friend.

Take a peek at what other nurseries charge for similar trees. However, please don't rush to undercut their prices as your first instinct. Instead, use this information to differentiate your offerings and create a unique value proposition.

Start by identifying what sets your tree nursery apart from the competition. Is it exceptional customer service, a wide selection of rare tree varieties, or specialized expertise in tree care? Highlight and showcase these strengths to potential customers to demonstrate why your nursery stands out from the competition.

When customers see the added value in your products and services, it makes them more willing to pay a premium. Emphasize the benefits of purchasing from your nursery, whether it's the health and quality of your plants, expert advice, or the overall experience they will receive.

Monitor seasonal trends

Keep an eye on seasonal trends and set/adjust your prices accordingly.

During peak planting seasons, demand for some tree species may skyrocket as customers gear up for landscaping and gardening projects. You can slightly increase your prices during these periods to capitalize on the increased demand. However, when demand might dip during slower periods, it's a great opportunity to offer promotions and discounts to keep sales booming. Consider running limited-time offers, bundle deals, or special packages to entice customers and maintain steady sales.

Take advantage of holidays and special occasions as well. For example, during Arbor Day or Earth Day, you can run themed promotions that celebrate nature and the importance of planting trees. Offering discounts or incentives during these times can attract environmentally-conscious customers and boost sales.

Flexibility is vital when it comes to seasonal pricing. Keep a close eye on market trends and customer behavior during different times of the year. If you notice certain tree varieties are in high demand during specific seasons, adjust your prices to maximize profits.

Furthermore, consider the competitive landscape during peak seasons. When other nurseries raise their prices, it may be an opportunity to do the same while remaining competitive. However, ensure your pricing strategy aligns with the perceived value and unique offerings you provide to customers.

Remember, pricing is not a one-size-fits-all approach. It requires ongoing analysis and adaptation to meet the needs and expectations of your customers throughout the year. By being attentive to seasonal trends and offering promotions strategically, you can ensure your tree nursery remains a sought-after destination for customers all year round.

Monitor and analyze

Last but not least, regularly monitor and analyze your pricing strategy. Regularly monitoring and analyzing your pricing strategy is vital for the long-term success of your tree nursery business. It lets you stay responsive to market changes, customer preferences, and profitability.

Keep a close eye on sales data, customer feedback, and profit margins to gain valuable insights into the effectiveness of your pricing approach. By tracking sales data, you can identify trends in customer behavior, such as which tree varieties are the most popular or which pricing promotions lead to increased sales. This information helps you make informed decisions about adjusting prices for specific products or seasons.

Customer feedback is equally important. Listen to what your customers say about your pricing and adjust based on their suggestions or concerns. Positive feedback may indicate that your pricing is competitive and offers good value, while negative feedback could signal that adjustments are needed to meet customer expectations.

Monitoring profit margins ensures your pricing strategy is sustainable and profitable. It helps you understand if your pricing covers all costs, including production, labor, and overheads, while still generating a reasonable profit. If profit margins are too low, consider optimizing costs or reevaluating you're pricing. Fine-tuning your pricing approach based on data-driven insights ensures that your tree nursery remains competitive and financially viable. Don't be afraid to experiment with different pricing strategies, such as tiered pricing or bundle deals, to see what resonates best with your customers.

Remember, pricing is not a set-it-and-forget-it process. The market is dynamic, and customer preferences may change over time. Regularly reassess and adapt your pricing strategy to maintain a healthy balance between customer satisfaction and profitability. By doing so, you can keep your tree nursery thriving and continue to meet the needs of your valued customers.

Nursery owner's tips:

While the factors mentioned above will work, there are some elements you need to remember.

- One of them is presentation. Presentation is a crucial aspect of running a successful tree nursery. A well-organized and visually appealing nursery can elevate the perceived value of your trees and create a positive impression on customers.

  o Start by arranging your trees in a neat and organized manner. Group them by species, size, or type to make it easier for customers to find what they want. Use clear signage to label each section and to provide essential information about the trees and their characteristics.

  o Consider creating attractive displays that showcase your trees' beauty and unique features. Use eye-catching visuals, like vibrant colors and artistic arrangements, to draw attention to specific species or seasonal highlights.

- Maintain a clean and tidy nursery environment: Regularly remove dead leaves, weeds, and debris to create a welcoming atmosphere that invites customers. A well-maintained nursery reflects your commitment to quality and attention to detail.

  o Invest in proper lighting to showcase your trees, especially during darker hours or overcast days. Adequate lighting can enhance the beauty of your trees and make them more appealing to potential buyers. Incorporate elements of landscaping and garden design within your nursery.

  o Showcase mature trees in outdoor settings, allowing customers to envision how they would look in their landscapes. This can inspire them to purchase and add a personal touch to their outdoor spaces.

  o Provide informative signage and labels for each tree by offering details about its growth habits, care requirements, and ideal planting conditions. This information empowers customers to make well-informed decisions and builds trust in your expertise.

- If you have nursery staff, don't forget the importance of customer service. Train your staff to be knowledgeable, approachable, and ready to assist customers with their inquiries and needs. A friendly and helpful attitude can leave a lasting positive impression on visitors.

  o Overall, a well-presented nursery creates a memorable and enjoyable experience for customers, encouraging them to return and recommend your business to others. By putting effort into presentation, you can maximize the appeal of your trees and create a thriving tree nursery that stands out in the market.

- Be flexible and monitor market trends, customer feedback, and your financial goals. Adjust your pricing strategy when needed to keep growing and succeeding.

- Don't forget to be transparent. Display your prices and give customers all the details about the value they'll get. Please have no surprises at checkout!

- Maintain accurate financial records by implementing a robust accounting system: track income, expenses, and transactions related to your tree nursery.

Consider using accounting software or hiring a professional accountant to ensure accuracy and facilitate tax preparation.

- Consistently record all financial transactions, including purchases, sales, and expenses. Keep organized records of invoices, receipts, and financial statements. Regularly reconcile your bank accounts to ensure accuracy.

- Familiarize yourself with tax deductions and credits relevant to your tree nursery business. These might include deductions for business expenses, equipment purchases, and qualified education expenses.

- Monitor your nursery's financial health by generating regular financial reports, such as profit and loss statements and balance sheets. These reports provide insights into your business's performance and help you make informed decisions.

- Insurance also plays a critical role in protecting your tree nursery business from various risks and unforeseen events. So, ensure you insure your nursery business. Essential Insurance for a US tree nursery business includes:

- General Liability, Property, Workers' Compensation, and Commercial Auto cover various risks.

- Crop Insurance protects against weather and pest threats.

- Business Interruption and Product Liability ensure continuity and safety.

Prioritize these to ensure comprehensive coverage for your nursery's success and security.

WOW! WOW! Look at how far you've come, and you've been attentive and learning. Please take a minute to reflect on how far you've come and appreciate it because you are now ready to implement everything and create a thriving tree nursery!

Conclusion

As we conclude, you've embarked on a journey to create a flourishing tree nursery. From the first seed to the final sale, you've learned the art and science of nurturing life. Armed with knowledge and dedication, you're ready to cultivate trees and dreams.

May your nursery grow and thrive into a testament to your passion and perseverance. The world awaits the beauty you'll bring forth, one tree at a time.

All the best, fellow plantsman!

PS: I'd like your feedback.

If you are happy with this book, please leave a review on Amazon.

Please leave a review for this book on Amazon by visiting the page below:

https://amzn.to/2VMR5qr

www.ingramcontent.com/pod-product-compliance
Lightning Source LLC
Chambersburg PA
CBHW072216290526
45794CB00004B/1770